# THE U.S.S. *CONSTITUTION*

## HISTORIC LANDMARKS

Jason Cooper

The Rourke Book Company, Inc.
Vero Beach, Florida 32964

PHOTO CREDITS:
All photos © Susan Cole Kelly except pages 10 & 12 © James P. Rowan

PRODUCED & DESIGNED by East Coast Studios
eastcoaststudios.com

EDITORIAL SERVICES:
Janice L. Smith for Penworthy Learning Systems

**Library of Congress Cataloging-in-Publication Data**

Cooper, Jason, 1942-
    The U.S.S. Constitution / Jason Cooper.
        p. cm. — (Historic Landmarks)
    Includes index.
    ISBN 1-55916-329-1
    1. Constitution (Frigate)—Juvenile literature. [1. Constitution (Frigate)] I. Title.

VA65.C7 C66  2000
359.3'22'0973—dc21
                                                00–030589

**Printed in the USA**

# TABLE OF CONTENTS

# THE U.S.S. *CONSTITUTION*

The War of 1812 (1812-1815) between Great Britain and the United States ended long ago. Amazingly, however, the U.S.S. *Constitution* remains sturdy and afloat. The *Constitution* is an American warship that fought in the War of 1812.

Built in 1797, the *Constitution* is tied to a dock at the old Charlestown Navy Yard near Boston, Massachusetts. The yard is part of the Boston National Historical Park.

*The* Constitution *rests dockside at old Charlestown Navy Yard in Charlestown, Massachusetts.*

Because of its age and wartime history, the three-masted *Constitution* is a national treasure. The U.S. Navy keeps the ship **seaworthy** (SEE wur thee). In fact, the *Constitution* is the oldest **commisioned** (kuh MIH shund) warship still afloat. That means she has a crew of U.S. Navy sailors and she can sail.

A tugboat takes the old **frigate** (FRIH gut) for harbor cruises about three times a year. The *Constitution* has also made brief sails on her own, the last in 1997.

*Here in 1997 the* Constitution *made a voyage in Massachusetts Bay under her own power to celebrate her 200th birthday and the completion of four years of repair. Her last voyage by herself had been in 1881!*

Keeping an old, wooden sailing ship in tip-top condition is a big job. Wood rots from its contact with air and rain. Throughout her lifetime, the *Constitution* has had her rotting timbers replaced with new ones. Still, about 20 percent of the ship is original oak.

Frigates like the *Constitution* were built to be fast attack ships. They were light, quick, and well-armed with cannons. Frigates could overcome smaller ships and battle with other frigates. Their best defense against larger warships, however, was to sail quickly away.

*A U.S. Navy sailor in early 19th century uniform leads a* Constitution *tour.*

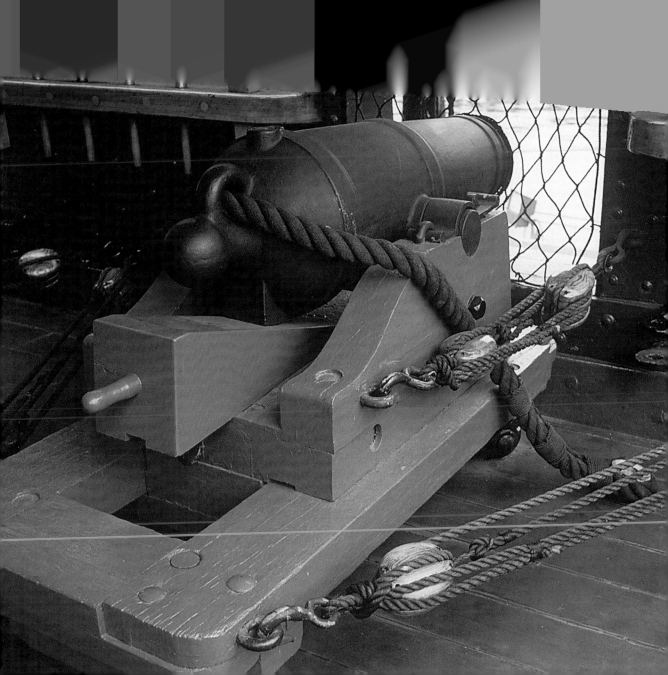

## DESIGNING THE *CONSTITUTION*

The *Constitution* was designed by Joshua Humphreys of Philadelphia in the 1790s. The U.S. Navy needed new warships to challenge Great Britain's Royal Navy. British ships ruled the seas. The British ships had been stopping American trade ships and sometimes taking seamen from them.

Humphreys designed the *Constitution* and her sister ships to be the largest, fastest frigates in the world. His design made the *Constitution's* undersides sleek. The streamlined shape gave her speed, up to 16 miles (26 kilometers) per hour with only wind power.

*This cannon fired 32-pound (14.5-kilogram) cannonballs, unusually heavy firepower for a frigate.*

*A sailor in 1812 uniform stands by one of the* Constitution's *24-pound cannons.*

*The Constitution's fighting sails are controlled by hundreds of yards of rope and a skilled U.S. Navy crew.*

Humphreys also made sure the American frigate would carry more and bigger guns than the British frigates. Sometimes the *Constitution* carried 50 guns on her decks.They fired 24- and 32-pound (11- and 14.5-kilogram) cannonballs.

The *Constitution* carried a crew of 450. They could raise her clouds of sails in minutes.

*A tug escorts the sleek* Constitution *into the harbor. The Navy is not likely to risk letting the ship sail by herself anytime soon.*

## THE *CONSTITUTION* AFLOAT

The *Constitution* proved her speed in July, 1812, during the War of 1812. Five British ships sailed after her. The British ships gave up after 60 hours of trying to catch the *Constitution*.

The *Constitution* proved her strength during the war, too. The *Constitution* had an exceptionally thick, dense skeleton of oak wood. Her sides were 2 to 3 feet (62-92 centimeters) thick.

*In her fighting days, the* Constitution *sailed with six sails, the three largest rigged to her three masts.*

In 1812, the *Constitution* fought the British frigate *Guerriere*. Cannonballs fired from the *Guerriere* bounced off the *Constitution's* sides. "Her sides are made of iron," American sailors cried. From that day, the *Constitution* became known as "Old Ironsides". That was long before real iron-sided ships were built.

During the War of 1812, the *Constitution* captured or destroyed 14 **vessels** (VEH sulz), including 5 warships. After the war, the Navy used the *Constitution* for a variety of jobs. During the Civil War (1861-1865) she was a training ship for sailors.

Constitution's *thick oak timbers and her speed afloat made her a threat to Britain's frigates and merchant ships.*

## ABOARD THE *CONSTITUTION*

You can walk the *Constitution's* deck today just as American seamen did 200 years ago. The ship is open 7 days a week. Navy sailors begin guided, 30-minute tours at 9:30 a.m. each day. Tours end by 4 p.m.

*With her needlepoint bowsprit leading the way, Constitution sails away from her dock and past the Boston skyline.*

Sailors dress like U.S. Navy men of the early 1800s. In those days, the *Constitution* fired her guns at enemy targets. She still fires a cannon shot each day at sunset.

Visiting the *Constitution* is a great way to gain a sense of America's early naval history. You see the *Constitution's* guns, her sleeping **quarters** (KWAWR terz), and the captain's quarters. You can also see the amazing web of ropes that hoist and control the ship's sails.

# GLOSSARY

**commisioned** (kuh MIH shund) — having been put into
official service

**frigate** (FRIH gut) — a square-sailed warship

**quarters** (KWAWR terz) — a living area, especially aboard a ship

**seaworthy** (SEE wur thee) — having the ability to go easily
to sea

**vessel** (VEH sul) — a boat or ship

# INDEX

# FURTHER READING

Find out more about the U.S.S. Constituition and the War of 1812 with these helpful books and information sites:

Carter, Alden R. *The War of 1812: Second Fight for Independence*. Franklin Watts, 1993.

Boston Historical National Park
   *www.nps.gov/bost*
U.S.S. Constitution
   *www.history.navy.mil/constitution*